Mathestars
MENTAL
MATHS

WORKBOOK ^{Grade} 2

Om
KIDZ

An imprint of Om Books International

COLOURFUL KITES

Follow the code to colour the kites below.

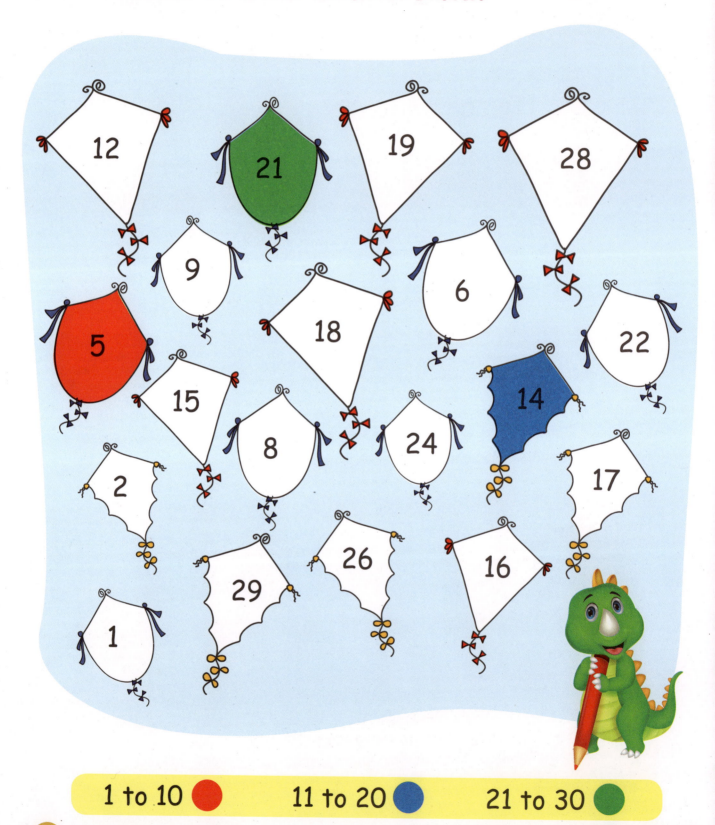

| 1 to 10 🔴 | 11 to 20 🔵 | 21 to 30 🟢 |

SPELL-O-COUNT

This Mathesaur is learning number names. Can you spell out the names of these numbers for him?

23 — Twenty-three

72 —

83 —

11 —

66 —

59 —

94 —

45 —

LOGGING AND SORTING

Can you arrange the numbers for the Mathesaur? Write them on the leaves in an ascending order. Write them on the stones in a descending order.

25 22 29 26 27

22 25 26 27 29

29 27 26 25 22

33 31 32 34 35

59 54 57 53 52

91 99 94 97 95

4

DINO DOODLE

Connect the numbers in the correct order. Then, colour the picture.

SHAPELY MATCH

Match the shape from column "A" to its name in column "B" for this Mathesaur.

A

B

cone

cylinder

sphere

pentagon

cube

cuboid

COLOURFUL FUN

Help the Mathesaur colour the 2D shapes red and 3D shapes blue.

3D TO 2D

Help the Mathesaur match the 3D shapes to their 2D lookalikes.

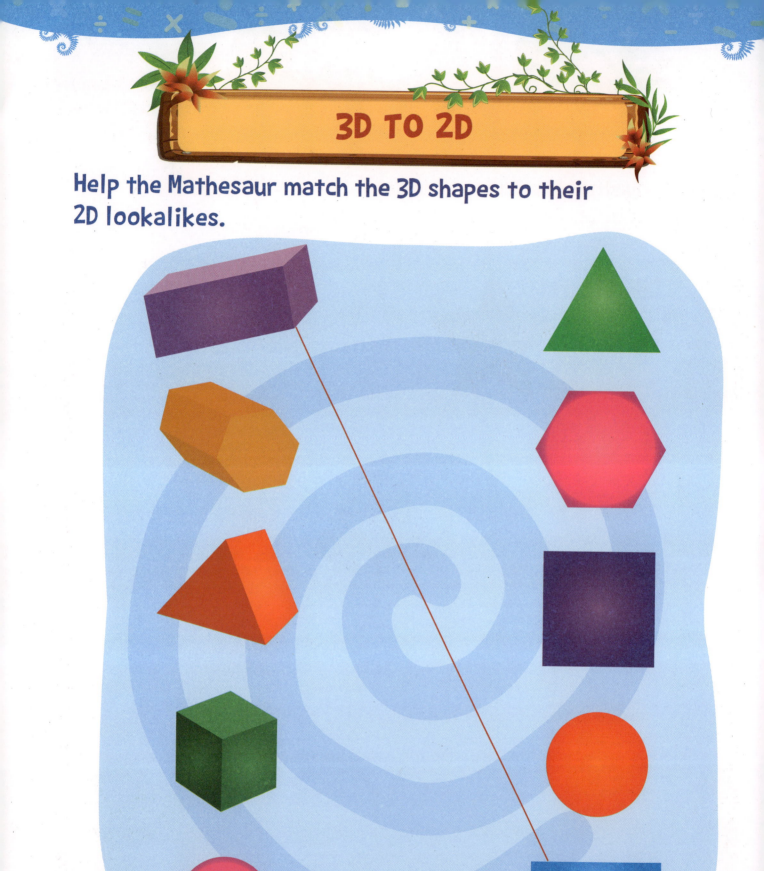

Help the Mathesaur write the names of the shapes below.

Triangle

ADDI-FUN

Let's solve some sums for the Mathesaur!

$$\begin{array}{r} 2 \\ +\ 2 \\ \hline 4 \end{array} \qquad \begin{array}{r} 4 \\ +\ 3 \\ \hline \end{array} \qquad \begin{array}{r} 5 \\ +\ 7 \\ \hline \end{array}$$

$$\begin{array}{r} 12 \\ +\ 3 \\ \hline \end{array} \qquad \begin{array}{r} 45 \\ +\ 7 \\ \hline \end{array} \qquad \begin{array}{r} 76 \\ +\ 9 \\ \hline \end{array}$$

$$\begin{array}{r} 88 \\ +\ 12 \\ \hline \end{array} \qquad \begin{array}{r} 34 \\ +\ 10 \\ \hline \end{array} \qquad \begin{array}{r} 65 \\ +\ 45 \\ \hline \end{array}$$

$$\begin{array}{r} 11 \\ +15 \\ \hline \end{array} \qquad \begin{array}{r} 68 \\ +15 \\ \hline \end{array} \qquad \begin{array}{r} 58 \\ +34 \\ \hline \end{array}$$

MATH MAZE

Follow the path that has sums of 20 and help the Mathesaur reach her egg.

CLOTH QUIZ

Given below are the clothes that this Mathesaur found.

Answer the following:

1. How many clothes did the Mathesaur find?_____20_____

2. How many red and pink t-shirts did he find?

_____

3. How many orange and green t-shirts did he find altogether?_____

SUBTRACTASAURS

Help the Mathesaur subtract the sums given below.

9 − 7 ——— 2	6 − 6 ———	7 − 3 ———
25 − 4 ———	76 − 5 ———	88 − 9 ———
55 − 45 ———	27 − 18 ———	92 − 73 ———

MARBLE MANIA

Subtract the marbles to solve the sums.

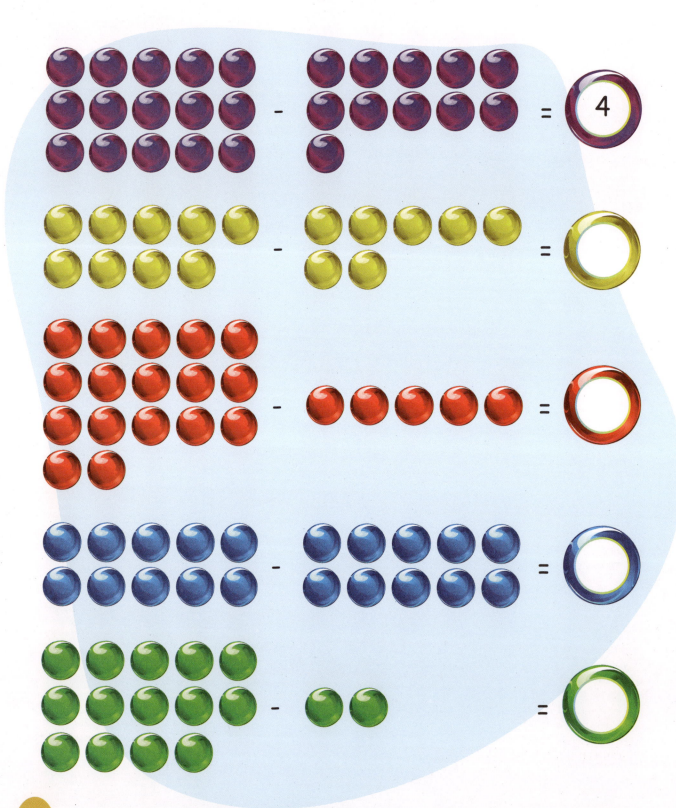

THE MATH BAKERY

This Mathesaur is starting a bakery!

Here are the cakes he has:

Here are the cakes he sold already:

Answer the following: 17

1. How many cakes does the bakery have in all?   ___

2. How many cherry cakes does it have in all?   ___

3. How many blueberry cakes does it have in all?   ___

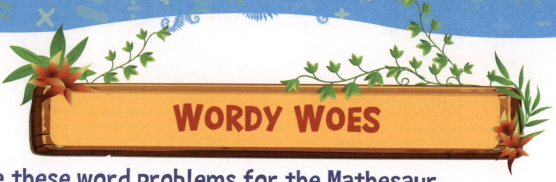

WORDY WOES

Solve these word problems for the Mathesaur.

1. Matt had 39 seashells. He gave 12 to his friend Fred and 8 to his friend Ted. How many seashells does Matt have now?_____19_____

2. Ryan ate 12 bananas for breakfast, 16 apples for lunch and 22 oranges for dinner. How many fruits did he eat in all?_____

3. Dylan had 45 marbles. He gave 18 to his sister, 12 to his brother and 5 to his mother. How many marbles are left with him?_____

4. Mitch gave 13 flowers to his mother, 12 flowers to his aunt and 19 flowers to his grandmother. How many flowers did he give in all?_____

5. Phil solved 23 sums yesterday and 54 sums today. How many sums did he solve in all? _____

FUN FAIR

These Mathesaurs are buying some toys at the fun fair! Look at the money they have and match them to the toy they can buy.

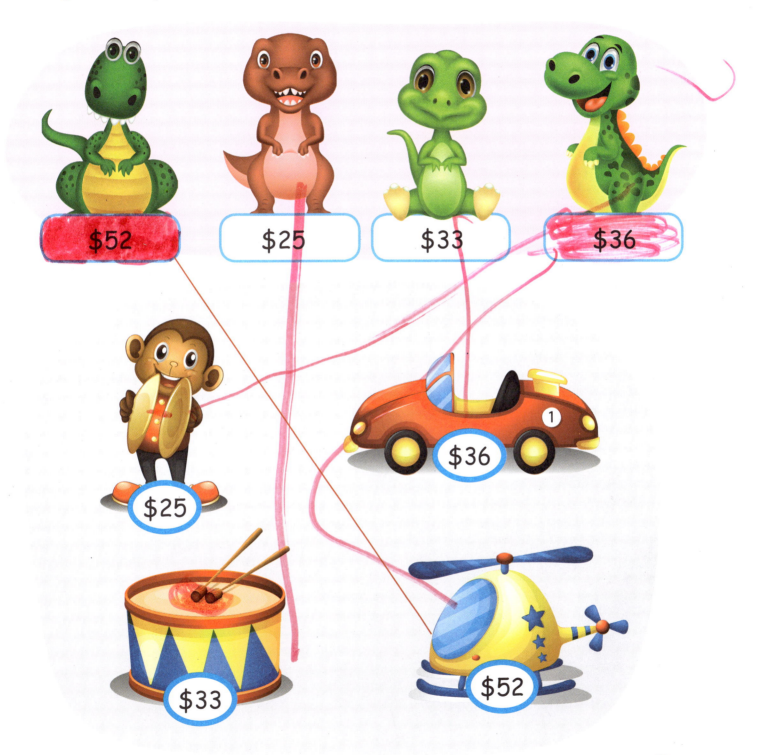

$52 $25 $33 $36

$25

$36   1

$33 $52

BIRTHDAY GIFTS

This Mathesaur has $50 in all. How many of these things can she buy within this money? Circle the pictures, but make sure that she doesn't spend more than $50 in all!

$5

$20

$60

$15

$10

HATTABOY!

This beautiful hat is on sale for $15. Tick the notes that the Mathesaur needs to buy it.

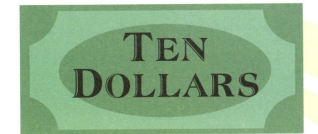

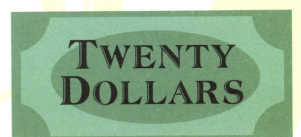

WEIGHTY WOES

Look at the balancing machine and answer these questions for the Mathesaur.

Which fruit weighs less?

_____Bananas_____

Do these foods have the same weight?

_____

Do these foods have the same weight?

_____

Which fruit weighs more?

_____

PUZZLING WEIGHTS

Solve the puzzle below.

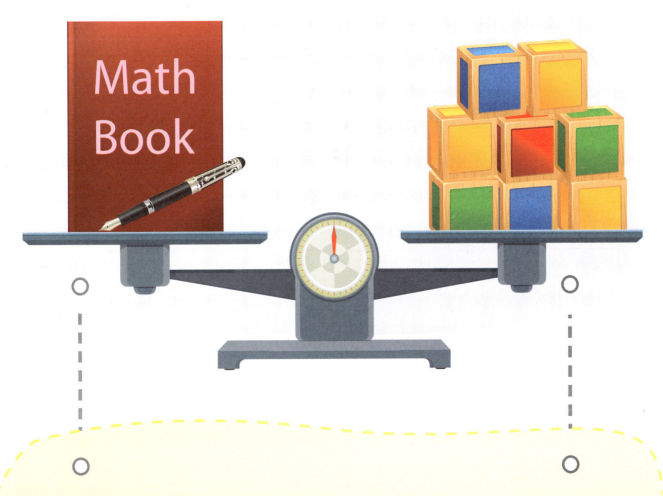

It takes 8 cubes to balance my book and pen together.

I know that it takes 5 cubes to balance my book by itself.

How many cubes does it take to balance my pen by itself?

_____

LENGTHY TRICKS

Help the Mathesaur circle the tallest one and underline the shortest one in each set.

CLOCK-A-DOODLE-DOO

Look at the time below each clock. Draw the hands on the clocks accordingly.

5:15

4:20

6:40

8:45

3:35

7:05

TIME-TELLER

Read the time on each clock. Write the time in the empty boxes below.

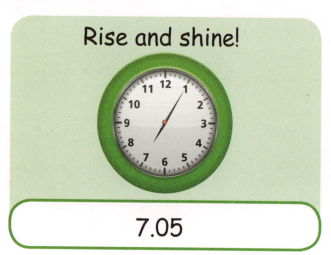

Rise and shine!

7.05

Let's leave for school.

Let's go home.

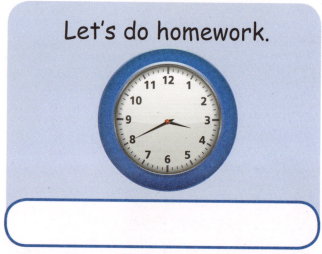

Let's do homework.

Let's play!

Let's say goodnight.

DIGI-WATCH

Look at the time shown on the digital stopwatch and help the Mathesaur to complete the sentences.

It took Amanda <u>three hours, thirty minutes and six seconds</u> to finish her homework.

Brown took _____ to have breakfast.

Kelly sleeps for _____ in the night.

Mike solves sums for _____ every evening.

MONTHLY WOES

There are twelve months in a year.
Write their names below.

Months in a year

January

MONTHLY MATCH

Help these Mathesaurs match the "month" leaves to the basket with the correct number of days.

HEX-A-PUZZLE

Each hexagon is made by adding the numbers in the two hexagons below it. Can you fill up the rest of the hexagons for the Mathesaur?

STICKY SOLVER

Can you remove 8 sticks from the figure below, such that the remaining sticks make 2 squares that do not touch each other. Help the Mathesaur solve the puzzle.

MATH HUNT

Help the Mathesaur by crossing out the line whose sum equals to the number written above the circles.

Make 8

4	6	5
3	2	1
3	1	2

Make 10

2	4	5
6	5	1
2	1	3

Make 12

5	1	3
2	6	4
3	2	5

Make 18

3	8	7
5	1	6
2	9	4

ANSWER KEY

Grade 2

Page 2
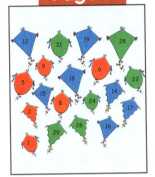

Page 3

- 23 Twenty-three
- 72 Seventy-two
- 83 Eighty-three
- 11 Eleven
- 66 Sixty-six
- 59 Fifty-nine
- 94 Ninety-four
- 45 Forty-five

Page 4

Page 5

Page 6
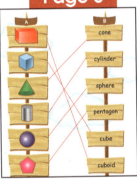

cone, cylinder, sphere, pentagon, cube, cuboid

Page 7

Page 8
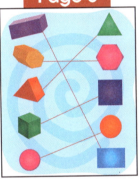

Page 9

Triangle, Rectangle, Cone, Sphere, Square, Cube

Page 10
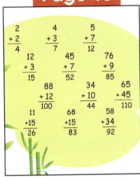

Page 11

Page 12
Answer the following:
1. How many clothes did the Mathesaur find? 20
2. How many red and pink t-shirts did he find? 5
3. How many orange and green t-shirts did he find altogether? 5

Page 13
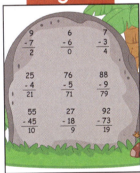

Page 14
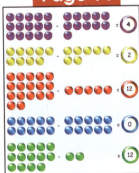

Page 15
Answer the following:
1. How many cakes does the bakery have in all? 17
2. How many cherry cakes does it have in all? 5
3. How many blueberry cakes does it have in all? 3

Page 16

1. Matt had 39 seashells. He gave 12 to his friend Fred and 8 to his friend Ted. How many seashells does Matt have now? 19
2. Ryan ate 12 bananas for breakfast, 16 apples for lunch and 22 oranges for dinner. How many fruits did he eat in all? 50
3. Dylan had 45 marbles. He gave 18 to his sister, 12 to his brother and 5 to his mother. How many marbles are left with him? 10
4. Mitch gave 13 flowers to his mother, 12 flowers to his aunt and 19 flowers to his grandmother. How many flowers did he give in all? 44
5. Phil solved 23 sums yesterday and 54 sums today. How many sums did he solve in all? 77

Page 17

ANSWER KEY

Page 18

Page 19

Page 20

Page 21

It takes 8 cubes to balance my book and pen together.
I know that it takes 5 cubes to balance my book by itself.
How many cubes does it take to balance the pen by itself?

3

Page 22

Page 23

Page 24

Page 25

03:30:06 — It took Amanda three hours, thirty minutes and six seconds to finish her homework.

01:05:20 — Brown took one hour, five minutes and twenty seconds to have breakfast.

08:30:20 — Kelly sleeps for eight hours, thirty minutes and twenty seconds in the night.

00:30:25 — Mike solves sums for thirty minutes and twenty-five seconds every evening.

Page 26

Page 27

Page 28
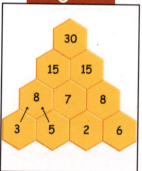

Page 29

Page 30